Antonio Rafael da Silva Mesquita
Elisabeth Regina Alves C. Silva

Temporal analysis of land use and occupation

Antonio Rafael da Silva Mesquita
Elisabeth Regina Alves C. Silva

Temporal analysis of land use and occupation

Case study in the Imbiribeira neighbourhood - Recife/PE

ScienciaScripts

Imprint

Cover image: www.ingimage.com

This book is a translation from the original published under ISBN 978-613-9-70232-9.

Publisher:
Sciencia Scripts
is a trademark of
Dodo Books Indian Ocean Ltd. and OmniScriptum S.R.L publishing group

120 High Road, East Finchley, London, N2 9ED, United Kingdom
Str. Armeneasca 28/1, office 1, Chisinau MD-2012, Republic of Moldova, Europe
Printed at: see last page
ISBN: 978-620-8-17500-9

I dedicate this monograph to my parents, who have always been present in my life in bad times and good times, relieving the burden I've carried throughout my journey.

ACKNOWLEDGEMENTS

I thank God first of all, because we can trust him at all times in our lives, and we are sure that he will be with us on this journey.

To my parents, Maysa Carvalho and Antonio Mesquita, who taught me to get up after stumbling and never give up on my goals. They taught me never to lower my head no matter how difficult the path I have to walk.

To my sister and my niece for putting up with me, because without them I wouldn't be able to forget the momentary problems, and for the affection exchanged during our time together.

To my girlfriend Daiana for her support and attention during difficult times, always encouraging me and giving me the strength not to give up, to my grandmother for always cheering for everyone and praying for God's protection.

To my family for always being there, to my friends for the fun and relaxation.

To the masters and doctors who taught us during the course, imparting invaluable knowledge that I will carry with me throughout my life, to the INBEC staff for their attention and promptness in providing services.

To my course mates who shared experiences that gave me a new outlook on life and to all the others who helped me along the way and were not mentioned, but who are no less important, thank you in advance!

"The cartographic base is essential... We, the users, must always check, evaluate and question the information contained on maps in order to ensure that their accuracy and content are continually improved and updated or minimally adapted to their use."

Valther Xavier Aguiar

SUMMARY

This work aims to elucidate the prerogatives relating to urban growth and the reduction of degraded vegetation as a result of subnormal urban development, where neighbourhoods are built without planning and without order, damaging the fauna and flora of the place where they are located, further reducing the strip of vegetation and also altering the use and coverage of the soil. In this sense, the neighbourhood chosen as a case study was Imbiribeira, because it is an area close to streams and bodies of water, where the lack of vegetation would have a greater impact because it acts as riparian forest, thus reducing anthropogenic interference, in addition to checking the entire vegetation strip located within the neighbourhood in question, verifying the green percentage, which also serves to reduce the temperature of the region's microclimate. Other ranges of land use and occupation were also stratified, showing the nuances in the years 1984 to 2016 between ranges of images acquired from the INPE website, where they were georeferenced with the help of an orthorectified image, showing the advances of the urban area, which is a characteristic present in metropolitan regions, to the detriment of the vegetation areas and exposed soil of the region analysed.

Keywords: Georeferencing. Vegetation. Land cover.

SUMMARY

CHAPTER 1

INTRODUCTION

Population growth is characterised as an increase in the number of inhabitants on the planet. This phenomenon is a consequence of vegetative growth, obtained through the balance between birth rates (births) and death rates (deaths). When the birth rate is higher than the death rate, we have positive vegetative growth, otherwise the rate is negative.

It wasn't until the end of the 17th century and the beginning of the 18th century that the world's population growth intensified, since before that time life expectancy was very low, which increased mortality rates. In 1930, the Earth was inhabited by around 2 billion people and, by 1960, this figure had reached the 3 billion mark, with an average population growth rate of 2% per year. During the 1980s, the world's population surpassed the 5 billion mark.

According to data released in 2010 by the United Nations Population Fund (UNFPA), the world's population is 6.908 billion. (FRANCISCO, W. de C.) The process of urbanisation in Brazil began at the end of the 19th century, when industry became the most important sector for the economy, and intensified in the middle of the last century, when it took on an air of modernisation. After the urban revolution of the 1950s, medium-sized cities multiplied, which, accompanied by the process of metropolisation, led to a considerable increase in large cities. With the accelerated growth of cities, which is generally not accompanied by the act of planning their occupation and expansion, certain urban problems have been potentiated and have acquired a character of environmental vulnerability. The expansion of urban space has led to increased pressure on areas and systems under environmental threat (BARCELLOS; OLIVEIRA, 2008.p. 3).

This population growth and urban sprawl means that the new inhabitants are looking for a place to live. These places are not always planned, which causes disorganised urban growth, often due to a lack of public administration, When looking for space in the already limited and saturated urban area, the population ends up invading the spaces destined for the preservation of vegetation, which in turn has the function of preserving the bodies of water that intersect urban regions, thus avoiding silting and contributing to the reduction of impurities that are dumped into rivers every day due to the lack of basic sanitation.

The current stage of metropolitan growth is characterised by the importance assumed by the environmental dimension of urban problems, especially those associated with the subdivision, use and occupation of land, with a relevant role played by housing settlements for low-income populations. (Grostein M. D.)

In recent years, human intervention in previously stable habitats has increased significantly, generating greater losses of biodiversity. Biomes are being occupied at different scales and speeds: extensive areas of native vegetation have been devastated in the Cerrado of Central Brazil, the Caatinga and the Atlantic Rainforest. (MMA)

As a result of the pressures of human occupation in the coastal zone, the Atlantic Rainforest, for example, has been reduced to approximately 7 per cent of its original vegetation.

In Recife there are several strips of vegetation protected by law from the Atlantic Forest, such as the APA Campo do Jiquiá and the Parque Natural Municipal dos Manguezais, the latter located in the neighbourhoods of Pina, Boa Viagem and Imbiribeira. The Imbiribeira neighbourhood has a total population of 48.512 inhabitants (2010 Census), we have a territory of 666 ha bordering two rivers (Rio Pina and Rio Tejipió) and housing the city's only natural lagoon, the Araçá lagoon, so this is the neighbourhood to study the vegetation strips in different years and the impact this has on the substantial increase in population in the highlighted region, thus knowing the interactions with the vegetation strips, bodies of water and exposed soil due to the unbridled and unplanned growth of low-income populations in the region.

In order to identify what urban pressure is doing to the vegetation strips located in the Imbiribeira neighbourhood, several different temporal images were stratified, with the aim of identifying and quantifying this area, and the percentage (%) that each attribute recognised in the images represents.

1.1 Objectives

1.1.10 General Objective

Showing the evolution of urban growth using image processing techniques in the Imbiribeira neighbourhood.

1.1.11 Specific Objective

Show how urban growth influences the region's vegetation strips, exposed soil and water bodies using techniques to verify land use and occupation, highlighting vegetation in this process.

1.2 Methodology

The methodology used during this work was to qualify the images captured on Google and on geoprocessing sites that make them available, such as INPE (National Institute for Space Research), transmitting individual values for each colour band representing a different layer, resulting in a thematic matrix map, using the ENVI programme made available by the Brazilian Institute for Continuing Education (INBEC) and ArcGIS software, where the unsupervised classification method was used in this specific composition with the Iso Cluster method, the temporal images of the imbiribeira neighbourhood were processed, verifying and stratifying the areas with vegetation, not identifying native or exotic, uniting them in a single layer, also verifying the urban area of the neighbourhood, areas of wasteland and strips of water bodies, thus verifying the use and occupation of the land, these strips were quantified and added up, reaching a maximum percentage and then being stratified so that there was knowledge of the amount of each strip separated by the software. The years used in this methodological quantification were 1984, 1998, 2008 and 2016.

CHAPTER 2

HISTORY OF THE CITY OF RECIFE

Its history began in 1534 along with the hereditary captaincies. The city was used as a harbour to load local produce and receive others from the metropolis. Its name was given by the large number of reefs on the region's coastline.

In 1630, the Dutch came to Brazil attracted by sugar cane and settled in Recife. Seven years later, Count Maurício de Nassau took over the government and developed the city. He planned and laid out streets, built bridges and brought architects, engineers and landscapers from Europe to improve the city's appearance. You can see the architecture of these people in the buildings still standing in the city.

In 1654, when the Dutch were expelled from Recife, the city was already a trading post, so the mill owners moved to Olinda. The rivalry between the cities, which began when Olinda was the capital of Pernambuco and when the Dutch preferred Recife to the capital, led to the Mascats' War. By 1848, the city was already a commercial and cultural hub for the entire Northeast region and soon became a distribution centre.

Figure 1 - Panorama of the city of Recife in 1855

Author: Friedrich Hagedorn

In the 20th century, Recife was boosted by large industries that settled in its centres, becoming the city's biggest activity. Today, Recife is the second largest medical centre in the country and one of the cities most visited by tourists. Recife has adapted its infrastructure to receive people arriving from all over the country and beyond. (Brasil Escola)

The city of Recife enters the 21st century consolidated as a major metropolis, with an area of 218,435 square kilometres and an estimated population of 1,617,183, classified by the Brazilian Institute of Geography and Statistics (IBGE) as a regional metropolis. (Visit Recife)

CHAPTER 3

CHARACTERISTICS OF THE CITY OF RECIFE

Recife's territory is unique. Gentle hills to the north, south and west structure an amphitheatre that shelters a plain cut by rivers and bathed by the sea, constituting the environment on which the city was built. This physical-geographical support, covered with remnants of the Atlantic Forest and its associated ecosystems, was gradually occupied, transformed and, in some places and at times, preserved - thus defining a landscape identity that characterises the intertwining of architecture and the natural site on which it was built.

According to Recife's Master Plan, Protected Units are those with mangrove forests, watercourses or bodies of water, as well as those of environmental or landscape interest necessary for the preservation of climate mitigation conditions, intended for recreational, sporting, social or leisure activities.

The city of Recife is currently home to a total of twenty-five Conservation Units, according to the Masterplan Figure (Environmental Sustainability Sectors) and the List of Conservation Units, which have been established by the Municipal Government since the 1996 Land Use and Occupation Law. Of these, five have already been categorised as Environmental Protection Areas (APA), according to the National System of Nature Conservation Units (SNUC - Federal Law No. 9.985/2000).

3.1 Geography

Recife is the capital of Brazil's seventh most populous state, Pernambuco, located near the parallel 8°04'03" south and the meridian

34°55'00" west. It covers an area of 218,435 square kilometres and is bordered by the municipalities of Jaboatão dos Guararapes, São Lourenço da Mata, Camaragibe, Paulista and Olinda.

Home to the Recife Metropolitan Region (RMR), the capital of Pernambuco has the fourth largest urban network in Brazil in terms of population, with a direct area of influence that covers the states of Alagoas, Paraíba, Pernambuco and Rio Grande do Norte (the latter together with Fortaleza), and also extends into the north of Bahia (together with Salvador).

3.2 Geology and relief

The city of Recife is situated on an alluvial plain (fluviomarine), made up of islands, peninsulas, wetlands and mangroves surrounded by five rivers: Beberibe, Capibaribe, Tejipió and the arms of the Jaboatão and Pirapama, giving it its peculiar characteristics. This plain is surrounded by arched hills that stretch from north to south, from Olinda to Jaboatão.

The average altitude in relation to sea level is four metres, but there are some areas of the municipality that are up to two metres below sea level, according to scholar Fernando Bruce. The municipality is located at latitude 8° 04' 03" S and longitude 34° 55' 00".

The city's boundaries are: to the south, hills of varying altitudes, such as Morro dos Guararapes, which stretch from Jaboatão dos Guararapes to Olinda; to the east, the coastline, which is lined with extensive sandstone ridges (reefs); and to the west, the municipalities of Camaragibe and São Lourenço da Mata.

3.3 Hydrography

Recife is known as the "Brazilian Venice" thanks to its similarity to the European city of Venice. Surrounded by rivers and crossed by bridges, it is full of islands and mangroves. The city is home to the Beberibe and Capibaribe rivers, which flow into the Atlantic Ocean. The town has dozens of bridges, including the oldest in Latin America, the Maurício de Nassau Bridge.

The vast network of rivers and canals in the Pernambuco metropolis includes the basins of the Capibaribe, Beberibe and Tejipió rivers and a drainage system made up of a series of **secondary watercourses or canals, tributaries or interconnected to the main drainage. Its water network is made up of watercourses such as the Moxotó, Jangadinha, Jiquiá and** Jordão **rivers**, as well as various canals, and it also receives contributions from the Capibaribe river through its **"dead arm", as well as significant bodies of water in areas such as *the* Araçá Lagoon, the** Várzea Dam and the Jangadinha Dam.

3.4 . Climate

Recife has a humid tropical climate (type As' in the Koppen-Geiger climate classification), typical of the northeastern east coast, with average monthly temperatures always above 18 °C, low thermal amplitudes and abundant rainfall throughout the year. The average annual temperature is 25.5 °C, reaching 30 °C in summer. Given the large number of skyscrapers in Recife, the formation of heat islands is common, which contributes to a difference in temperature between different regions of the city. According to a study by UFPE, in a single heat island in Recife the temperature can vary by up to 13 °C, and the neighbourhood most affected by this phenomenon is Boa Viagem.

Table 1 - Highest Rainfall Accumulations in Recife

Highest accumulations of rainfall in 24 hours recorded in Recife

(Curado, INMET) by month (03/1961-present)					
Month	**Cumulative**	**Date**	**Month**	**Cumulative**	**Date**
January	117 mm	07/01/2000	July	176.4 mm	29/07/1990
February	123.1 mm	16/02/1983	August	335.8 mm	11/08/1970
March	145.7 mm	19/03/2003	September	108.9 mm	17/09/2000
April	165.3 mm	22/04/1973	October	86.8 mm	14/10/1976
May	235 mm	24/05/1986	November	48.2 mm	24/11/1980
June	176.4 mm	12/06/1965	December	141.1 mm	06/12/2005

Source: National Metereological Institute

The average amount of sunshine is 2,550 hours per year, with relative humidity of 80%. Rainfall is over 2,000 millimetres (mm) per year, concentrated between April and July (monthly averages of over 300 mm), with July being the month with the highest rainfall (388 mm). Rainfall comes in the form of showers, which can be accompanied by lightning and thunderstorms and can be very intense, often causing flooding. The biggest flood in Recife's history was recorded in July 1975.

Snowfalls are more common in the wetter months, with one of the heaviest occurring on 30 October 1998, when the city dawned grey due to a thermal inversion. Although quite rare, cold fronts can also arrive. Windstorms are also uncommon, but one of the strongest occurred on 18 February 2010, when a gust of 86 km/h blew houses apart.

According to data from the National Meteorological Institute (INMET), since 1961 the lowest temperature recorded in the capital of

Pernambuco (Curado weather station) was 15 °C on 2 September 1965 and 4 August 1999, and the highest was 35.1 °C on 21 March 1988. The highest accumulated rainfall in 24 hours was 335.8 mm on 11 August 1970.

Some other large accumulations were 235 mm on 24 May 1986, 185.9 mm on 1 August 2000, 176.4 mm on 29 July 1990 and 12 June 1965, 165.3 mm on 22 April 1973, 162.8 mm on 29 June 1990, 162 mm on 21 July 1973, 159.7 mm on 10 June 1980 and 154.2 mm on 8 April 1986. The month with the highest rainfall was April 1973, when 770.4 mm were recorded. The lowest humidity level was recorded on 17 January 1998, at 41%.

CHAPTER 4

DIGITAL IMAGE PROCESSING

Digital image processing consists of performing mathematical operations on data in order to transform it into images with better spectral and spatial qualities and which are more appropriate for a given application. As such, image processing is individually designed for each particular type of problem. For example, a technique for accentuating the relief of a specific region will be ineffective for identifying the high frequencies that characterise changing patterns in urban areas.

Image processing is set up by specialised algorithms, which allow the user to apply a wide variety of processing techniques. The algorithms mathematically treat the image as data directly related to the physical process that created it. However, the representation and mathematical processing of the image is not 100 per cent a representation of the real world. In this situation, it is up to the user to analyse and interpret the data contained in the image, not just the computer.

4.1 Remote Sensing

By definition, the authors Paulo Roberto Menezes and Taita de Almeida (2008) inform us that the term Remote Sensing was created to designate the development of this new technology of instruments capable of obtaining images of the earth's surface at remote distances. For this reason, the best-known or classic definition of remote sensing is: Remote sensing is a technique for obtaining images of objects on the Earth's surface without any physical contact between the sensor and the object.

According to A. P. Crósta (1992), remote sensing images are

becoming more and more a part of our daily lives. Weather forecasts are illustrated in newspapers and on television by images from meteorological satellites, regions of international conflict are shown in the media using images taken from space, and satellite images are increasingly used to illustrate books, catalogues, calendars, posters and many other forms of visual communication.

From a technical-scientific point of view, remote sensing images have served as data sources for geological, agricultural, cartographic, forestry, urban and oceanographic studies and surveys, among others. Above all, remote sensing images have come to represent one of the only viable forms of environmental monitoring on local and global scales, due to their speed, efficiency, periodicity and synoptic view. At a time when humanity is beginning to take seriously the need to monitor the global changes taking place on the planet's surface, remote sensing appears to be one of the strategic tools for the future. (A. P. Crósta, 1992)

In view of the search for knowledge of human activity on the earth's surface, this tool is effective in diagnosing the causes and effects of anthropogenic activity around the earth.

4.2 Supervised and Unsupervised Classification

There are two types of image classification that could be tested during digital image processing work: Supervised and Unsupervised classification. These classifications differ in terms of knowledge of the study area, where the first would be with prior knowledge of the area and the second without this knowledge.

According to Crósta (1992), this type of classification requires the user to know something about the area to be classified before starting the process.

Ideally, there should be specific locations within the area to be classified where field observations were made during the satellite's passage to acquire the image. This type of observation is known as ground truth.

According to Crósta himself (1992), non-supervised classification is based on the principle that the computer (in reality, the software used) is capable of identifying the classes within a data set on its own.

The classification used in this work will be unsupervised, where they will be grouped by their spectral characteristics, organising them into clusters.

The parameters most commonly used to define clusters are the mean, variance and covariance, by which the spectral proximity of pixels in multidimensional space is defined. Clusters are therefore spectral classes.

This is an important difference from supervised classification, since in supervised classification it is up to the analyst to define a priori the classes they wish to separate (classes of interest), while in unsupervised classification the analyst only provides a few parameters such as the minimum and maximum number of classes desired and the number of iterations. The spectral classes generated in an unsupervised classification may or may not coincide with the classes of interest. It is therefore an exploratory operation to check what is statistically separable or not, which classes are the most obvious and their relationship with what you want to discriminate (IBGE, 1999).

In the Imbiribeira neighbourhood, 11 classes were used in the unsupervised classification to later form the four classes of interest (Vegetation, Urban Area, Exposed Soil, Water), as the images have a resolution of 30 x 30 metres, totalling 900 m^2 the pixel area, often making a pixel contain information from two of the classes mentioned, and so may contain divergences in both the supervised and unsupervised classifications.

4.3 Orthorectification

Orthorectification is a process that aims to correct an image so that each pixel is placed in the theoretical geometric position as if they had been imaged vertically or closer to it. In this way, orthorectification aims to generate an image in which internal (system) and external (landscape) distortions are corrected, making the coordinates more accurate, and correcting the effects of perspective and the influence of relief (valleys and hills) on the geometry of the image. (Martin, L., 2017)

All the images used had to be georeferenced because they were out of sync with an image provided by INBEC and with the layers downloaded from IBGE.

CHAPTER 5

STUDY AREA

The Recife City Council built the Recife Municipal Atlas in 2005, based on the Human Development Index (HDI), using local development indicators broken down by Political-Administrative Region (RPA), intra-urban Microregions and Human Development Unit (UDH), with the aim of broadening knowledge about social issues in the city of Recife and providing local support for the implementation and/or redirection of public policies and social programmes (PREFEITURA DO RECIFE, ATLAS DO DESENVOLVIMENTO HUMANO DO RECIFE, 2005).

This index, as has been emphasised, has been used for comparative studies of territorial inequalities on an international, national - between metropolitan capitals - and local scale - between the municipalities in the central core of the metropolitan agglomeration and the others on the periphery. On an intra-municipal scale, with greater detail, comparisons between the 62 Human Development Units (UDHs) reveal the inequalities experienced by the people of Recife in their areas of residence (BITOUN, 2005).

The study area is located in the Political Administrative Region VI (RPA6) of the city of Recife, which includes the neighbourhoods of: Boa Viagem; Brasília Teimosa; Imbiribeira; Ipsep; Pina; Ibura; Jordão; Cohab. The neighbourhood chosen for the study was Imbiribeira, as shown in figures 2, 3, 4 and 5. It is a neighbourhood in the south of the city of Recife, part of the city's 6ª Political-Administrative Region, bordering the neighbourhoods of Pina, Boa Viagem, Ipsep, Areias, Jiquiá and Afogados.

The neighbourhood is crossed from north to south by Avenida Marechal Mascarenhas de Morais.

Figure 2 - Imbiribeira neighbourhood with localisation image (2016)

Source: Google Earth

Figure 3 - Imbiribeira neighbourhood with panoramic view

Source: Pinterest.com

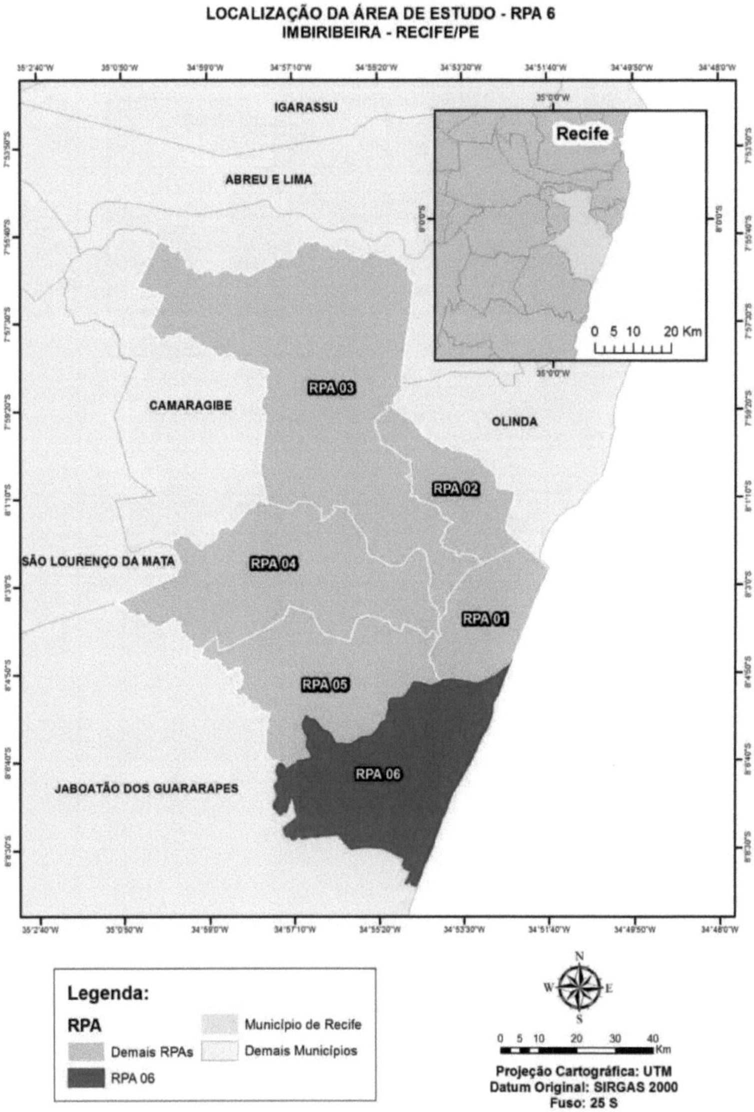

Figure 4 - Location of RPA 6

Source: Brazilian Institute of Geography and Statistics (IBGE); Elaboration: Author

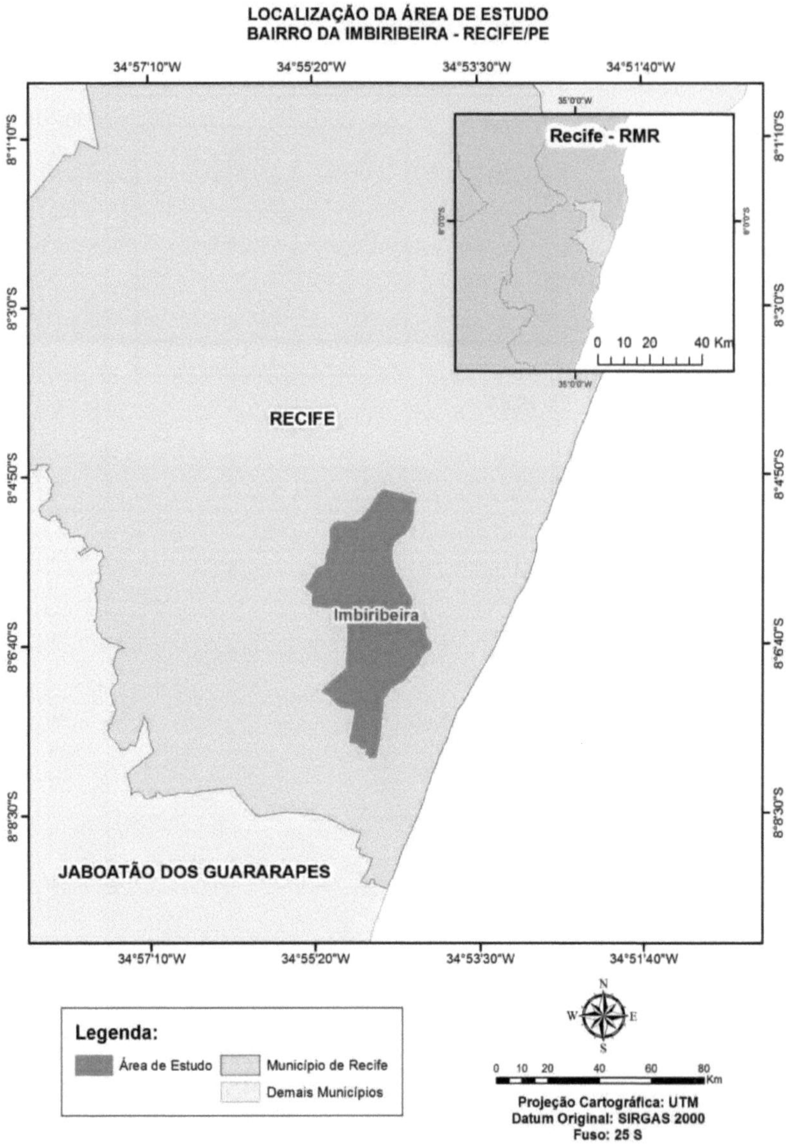

Figura 5 - Location of the Study Area
Source: Brazilian Institute of Geography and Statistics (IBGE); Elaboration: Author

The Imbiribeira neighbourhood was chosen for the study because it is a strategic environmental area. According to the Recife State Secretariat's stratification of protected areas, the Imbiribeira neighbourhood borders or aggregates at least four protected areas, as shown in figure 6.

LOCALIZAÇÃO DAS UNIDADES DE CONSERVAÇÃO MUNICIPAIS

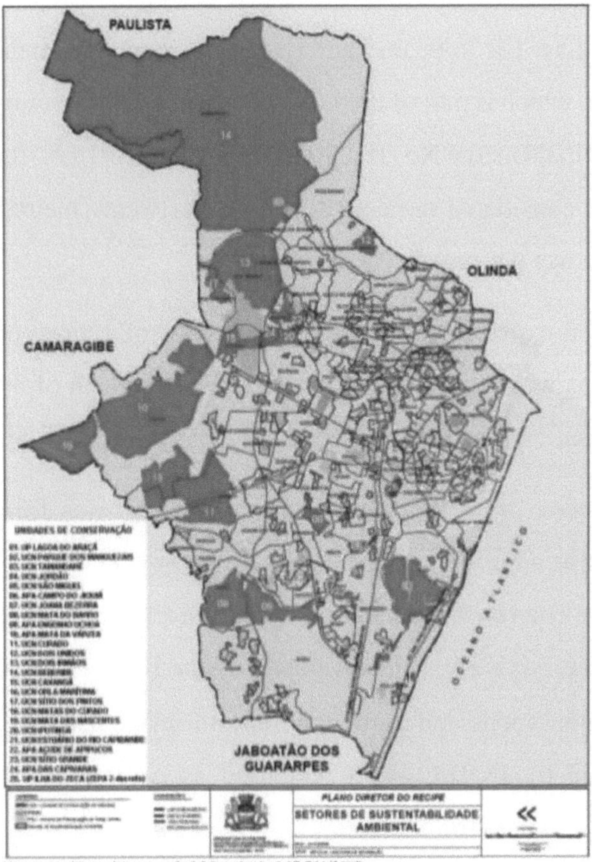

Figura do Plano Diretor – Lei Municipal nº 17.511/2008

Figura 6 - Location of Municipal Conservation Units

Source: Recife City Hall

According to the Recife Master Plan, the Environmental Sustainability Sectors in the Imbiribeira neighbourhood shown on the map are: 01 - UP Lagoa do Araçá; 02 - UCN Parque dos Manguezais; 04 - UCN Jordão; 23 - UCN Sitio Grande, together they cover more than 400ha and are home to various forms of wildlife, maintaining biodiversity and serving as a refuge for wildlife in an urbanised environment.

In accordance with Law 16176/96 on Land Use and Occupation on the banks of Lagoa do Araçá, which is part of the Lagoa do Araçá Conservation Unit, regulated by Municipal Decree No. 18.029 of 09 September 1998, the "non aedificandi" area is considered to be a strip of 20 m (twenty metres) away from the respective wet perimeters.

The images are then expected to detect several strips of vegetation during their processing, as well as bodies of water that border much of the neighbourhood in question.

The images were taken from the INPE website, the first was from 1984, where the image was adjusted using the ArcGIS geoprocessing tool in order to move it to the correct coordinates, as its orthorectification was not in congruence with the `layers` made available by the IBGE, such as the boundaries of neighbourhoods and municipalities, an image used as a base was made available by INBEC itself and used as a reference to georeference the images obtained from INPE.

The image retrieval process must be accompanied by a user registration, where the user receives information via links in their email, where they can download the images sent. An initial obstacle was choosing which temporal images to use, due to the amount of clouds that form on the Pernambuco coast, thus hindering the retrieval of images that can be used to carry out analyses, as exemplified in the figures below (All were taken from the INPE website):

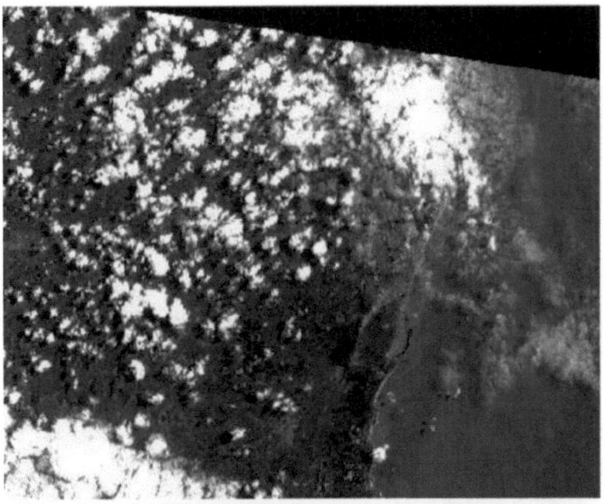

Figure 7 - 1987 Landsat image with clouds

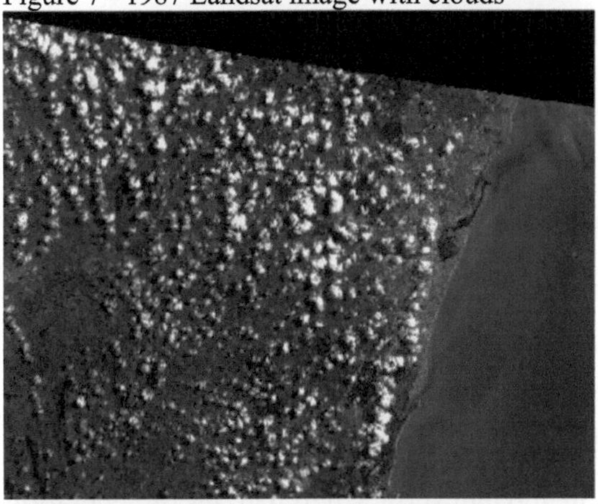

Figure 8 - 1995 Landsat image with clouds

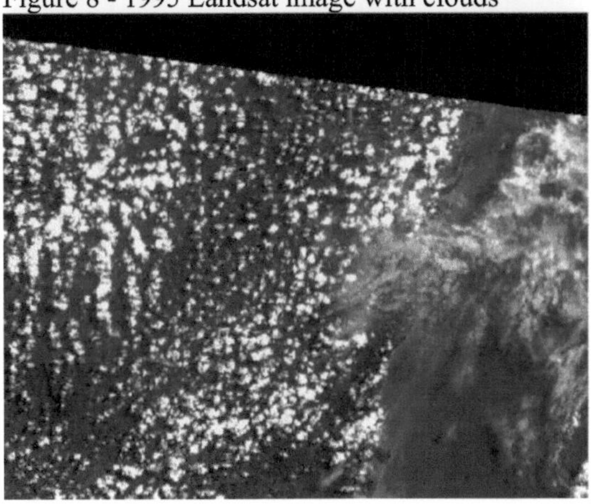

Figure 9 - 1996 Landsat image with clouds
Source:

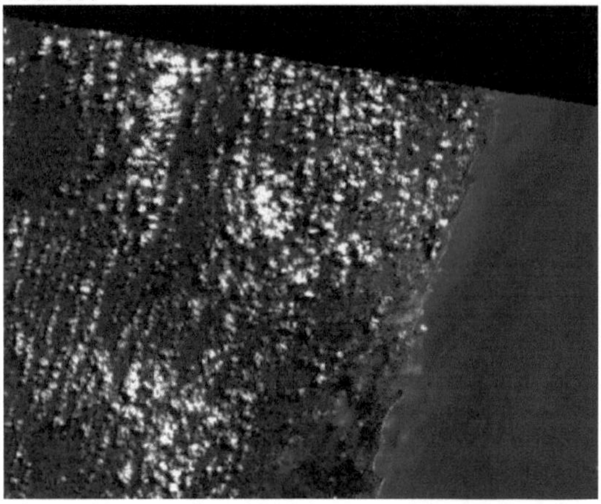

Figure 10 - Landsat 2000 image with clouds

Source

So the final images available for analysis and with the best possible visualisation and separation to identify the target bands were those from 1984, 1998, 2008 and a single one taken from Google Earth, quoted below.

The most up-to-date image (2016) was taken from google earth for the study, as it has the best resolution and can show small patches of vegetation where on a smaller scale it would be impossible to differentiate, for older images it was necessary to obtain an image with a lower quality in terms of resolution, due to satellites not having satisfactory image resolution technology made available to the civilian public, but they prove to be more efficient in the sense of image size, as the one obtained by google earth does not allow a larger zoom, distorting its characteristics due to its small size.

These images were processed by separating each value range found into a different category, where their content can be qualitatively analysed and the processed images can be judged on the basis of the classes found, thus providing an opinion on the evolution of urban growth that influences vegetation and generally changes the use and occupation of the land in the area studied. Initially, the supervised and unsupervised methods were used to demonstrate the effectiveness of each in obtaining land use and occupation classes, but only the data from the unsupervised method was shown in this research, which was stratified into 11 different symbologies and then separated into the bands intended for the study. Band 5 was used in all the satellite images, as shown in the figure below with the bands of the landsat 5 satellite and a description of the characteristics of the bands and their practical applications:

Banda	Intervalo espectral (µm)	Principais características e aplicações das bandas *TM* do satélite LANDSAT-5
1	(0,45 - 0,52)	Apresenta grande penetração em corpos de água, com elevada transparência, permitindo estudos batimétricos. Sofre absorção pela clorofila e pigmentos fotossintéticos auxiliares (carotenóides). Apresenta sensibilidade a plumas de fumaça oriundas de queimadas ou atividade industrial. Pode apresentar atenuação pela atmosfera.
2	(0,52 - 0,60)	Apresenta grande sensibilidade à presença de sedimentos em suspensão, possibilitando sua análise em termos de quantidade e qualidade. Boa penetração em corpos de água.
3	(0,63 - 0,69)	A vegetação verde, densa e uniforme, apresenta grande absorção, ficando escura, permitindo bom contraste entre as áreas ocupadas com vegetação (ex.: solo exposto, estradas e áreas urbanas). Apresenta bom contraste entre diferentes tipos de cobertura vegetal (ex.: campo, cerrado e floresta). Permite análise da variação litológica em regiões com pouca cobertura vegetal. Permite o mapeamento da drenagem através da visualização da mata galeria e entalhe dos cursos dos rios em regiões com pouca cobertura vegetal. É a banda mais utilizada para delimitar a mancha urbana, incluindo identificação de novos loteamentos. Permite a identificação de áreas agrícolas.
4	(0,76 - 0,90)	Os corpos de água absorvem muita energia nesta banda e ficam escuros, permitindo o mapeamento da rede de drenagem e delineamento de corpos de água. A vegetação verde, densa e uniforme, reflete muita energia nesta banda, aparecendo bem clara nas imagens. Apresenta sensibilidade à rugosidade da copa das florestas (dossel florestal). Apresenta sensibilidade à morfologia do terreno, permitindo a obtenção de informações sobre Geomorfologia, Solos e Geologia. Serve para análise e mapeamento de feições geológicas e estruturais. Serve para separar e mapear áreas ocupadas com *pinus* e *eucalipto*. Serve para mapear áreas ocupadas com vegetação que foram queimadas. Permite a visualização de áreas ocupadas com macrófitas aquáticas (ex.: aguapé). Permite a identificação de áreas agrícolas.
5	(1,55 - 1,75)	Apresenta sensibilidade ao teor de umidade das plantas, servindo para observar estresse na vegetação, causado por desequilíbrio hídrico. Esta banda sofre perturbações em caso de ocorrer excesso de chuva antes da obtenção da cena pelo satélite.
6	(10,4 - 12,5)	Apresenta sensibilidade aos fenômenos relativos aos contrastes térmicos, servindo para detectar propriedades termais de rochas, solos, vegetação e água.
7	(2,08 - 2,35)	Apresenta sensibilidade à morfologia do terreno, permitindo obter informações sobre Geomorfologia, Solos e Geologia. Esta banda serve para identificar minerais com íons hidroxilas. Potencialmente favorável à discriminação de produtos

Figure 11 - Landsat 5 Satellite Bands and their Characteristics and Applications
Source: Mundo Geo

For the Unsupervised Classification of the images, ArcGIS 10.3 was used as the data processing software.

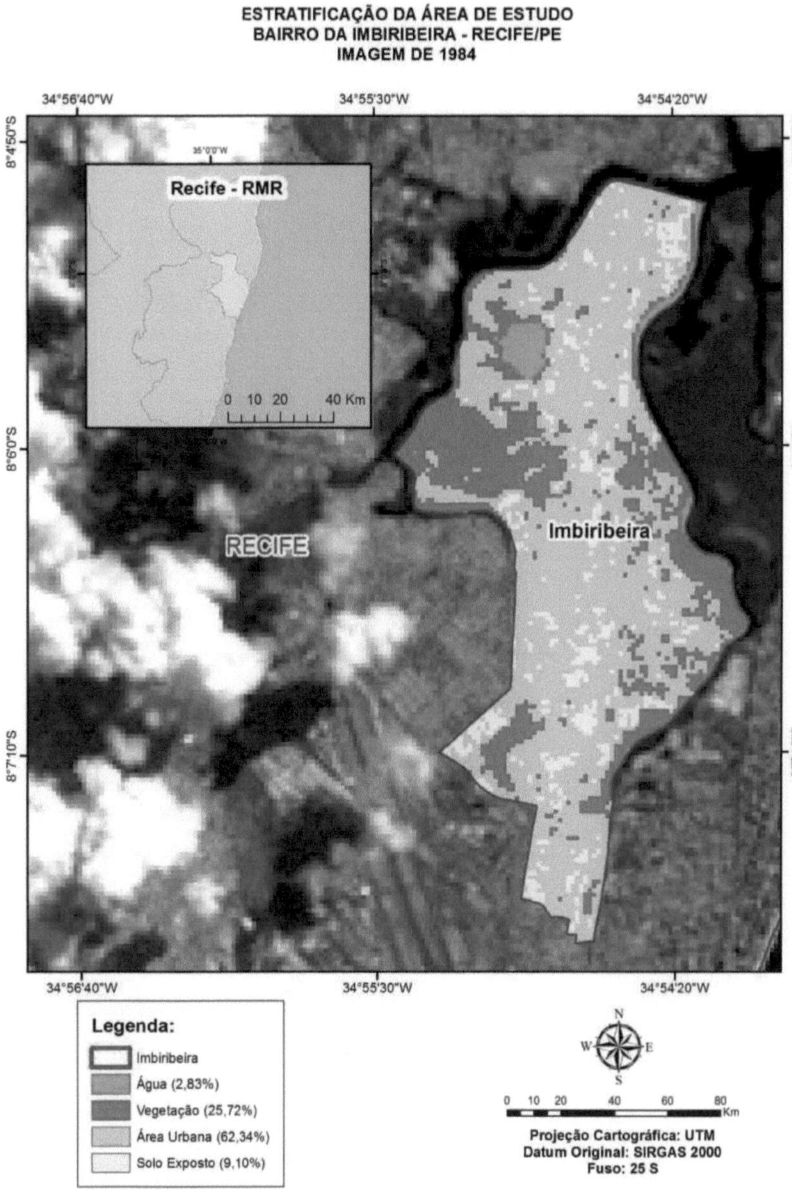

Figura 12 - **Imbiribeira neighbourhood with 1984 image**
Source: IBGE and INPE, Prepared by the author

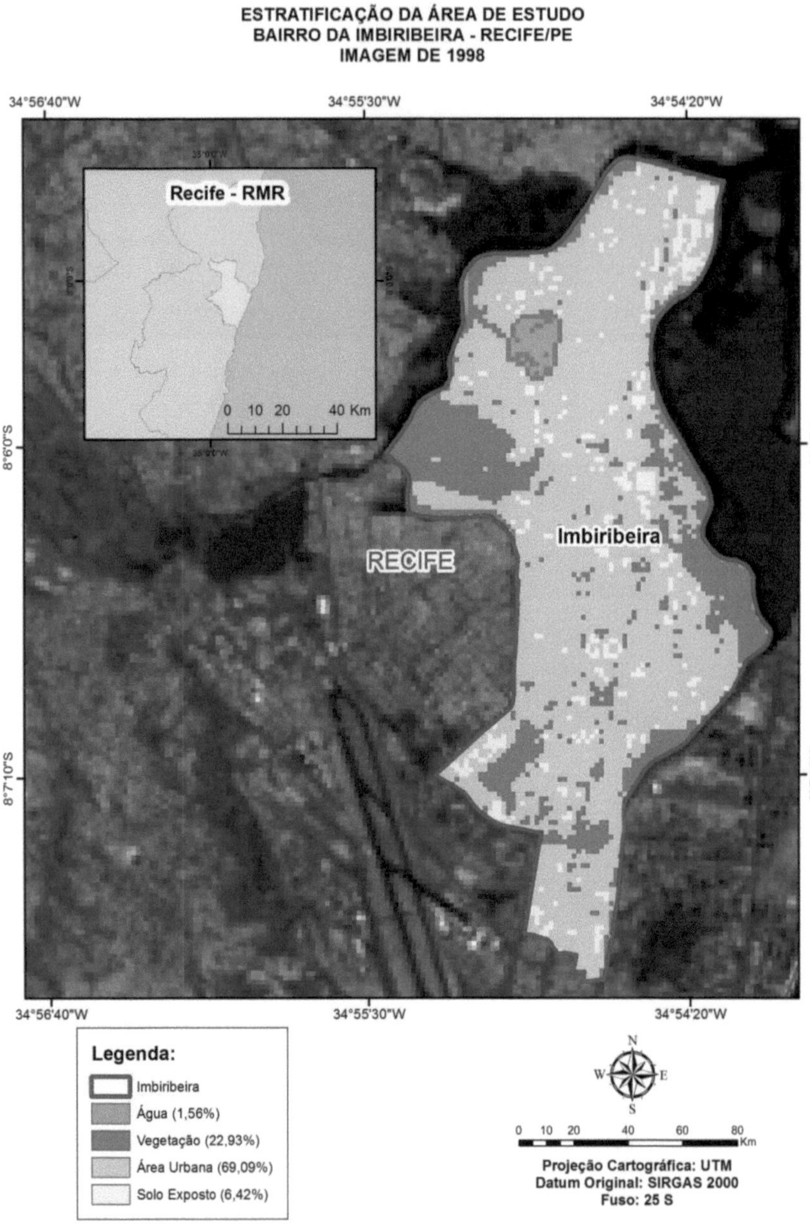

Figura 13 - **Imbiribeira neighbourhood with 1984 image**

Table 2 - Percentage Stratification of the Imbiribeira Neighbourhood - 1984

Stratification	%
Exposed Soil	9,10%
Water	2,83%
Urban Area	62,34 %
Vegetation	25,72 %
Grand Total	**100,00%**

In the images downloaded from INPE (Instituto de Pesquisas Espaciais) from 1984, a higher percentage of urban areas was identified, followed by vegetation, exposed soil came in third place and lastly, band 5 of landsat 7 was used in this image, as it had a greater segregation of vegetation and water from the other elements that make up the image, it had to be georeferenced with a tool from the ArcGIS software as it was displaced in relation to an orthorectified landsat image made available by INBEC and layers obtained from IBGE.

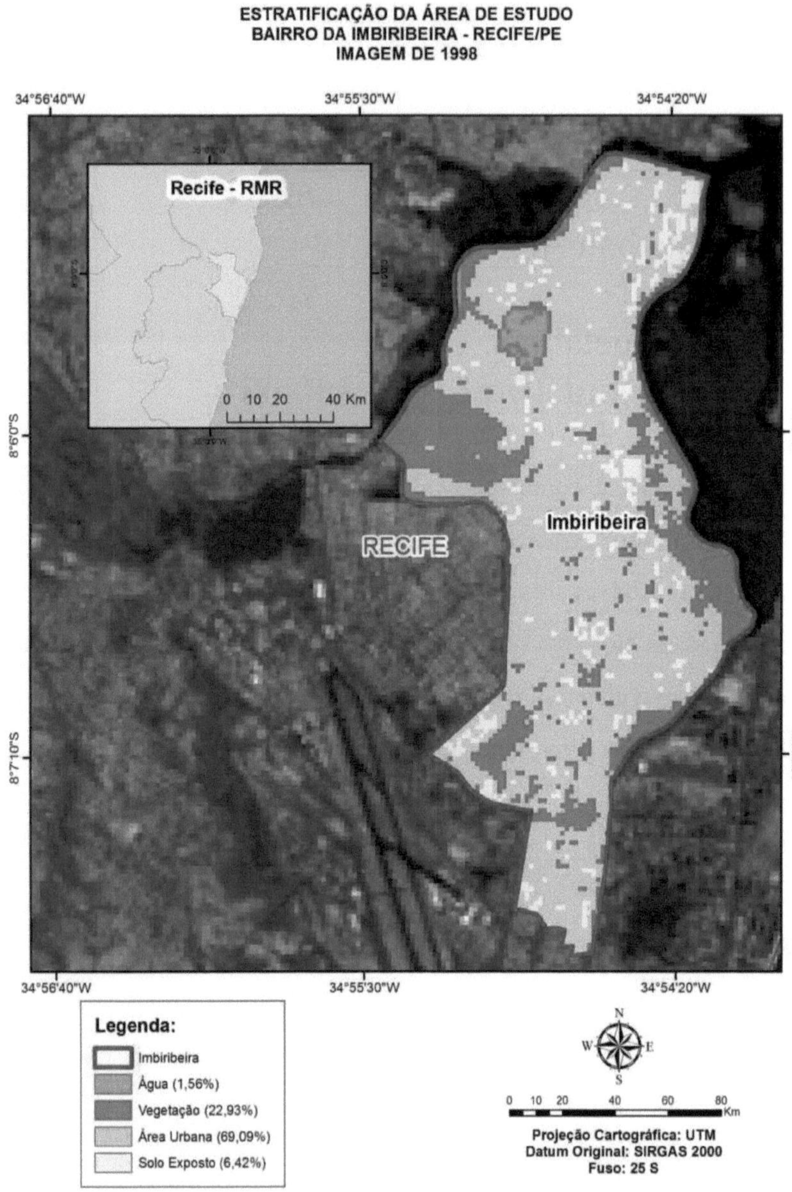

Figura 14 - **Imbiribeira neighbourhood with 1998 image**

Source: IBGE and INPE, Elaboration: Author

Table 3 - Percentage Stratification of the Imbiribeira Neighbourhood - 1998

Stratification	%
Exposed Soil	6,42%
Water	1,56%
Urban Area	69,09%
Vegetation	22,93%
Grand Total	**100,00%**

In the 1998 image, there was a decrease in the area of vegetation and exposed soil proportional to the increase in the urban area, totalling 69.09%. In this image, band 5 of the Landsat 7 satellite was used, where the vegetation is more evident; the image was also georeferenced in view of its displacement in relation to an orthorectified Landsat image made available by INBEC and layers obtained from IBGE.

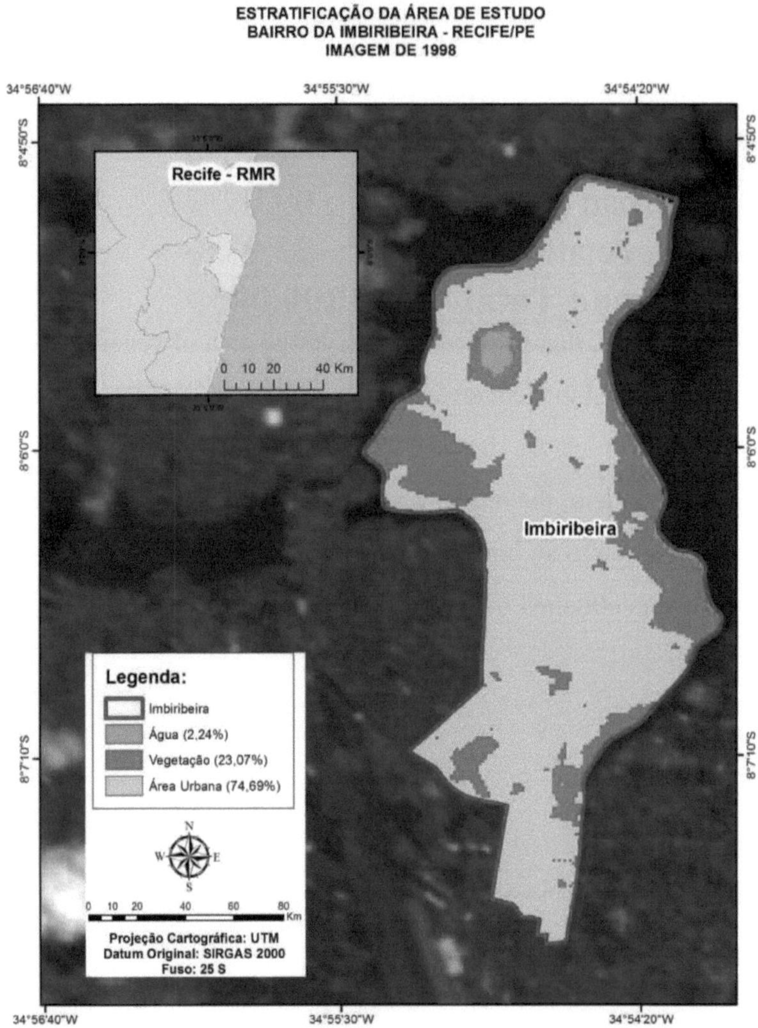

Figura 15 - **Imbiribeira neighbourhood with 2008 image**

Source: IBGE and INPE, Elaboration: Author

Table 4 - Percentage Stratification of the Imbiribeira Neighbourhood - 2008

Stratification	%
Exposed Soil	0%
Water	2,24%
Urban Area	74,69%
Vegetation	23,07%
Grand Total	**100,00%**

In the 2008 image there was an increase in the vegetation band, a decrease to 0% exposed soil proportional to the increase in the urban area band, totalling 74.69%. This image used band 2 of the CBERS satellite where both vegetation and bodies of water were more evident, and the image was also georeferenced in view of its displacement in relation to an orthorectified Landsat image made available by INBEC, A curiosity of this image is that it was not possible to define several areas of water bodies in differentiation with the vegetation, and the urban area with the symbolism of exposed soil, which is probably why the two symbols were merged, thus increasing the percentage of urban area to the detriment of the exposed soil layer.

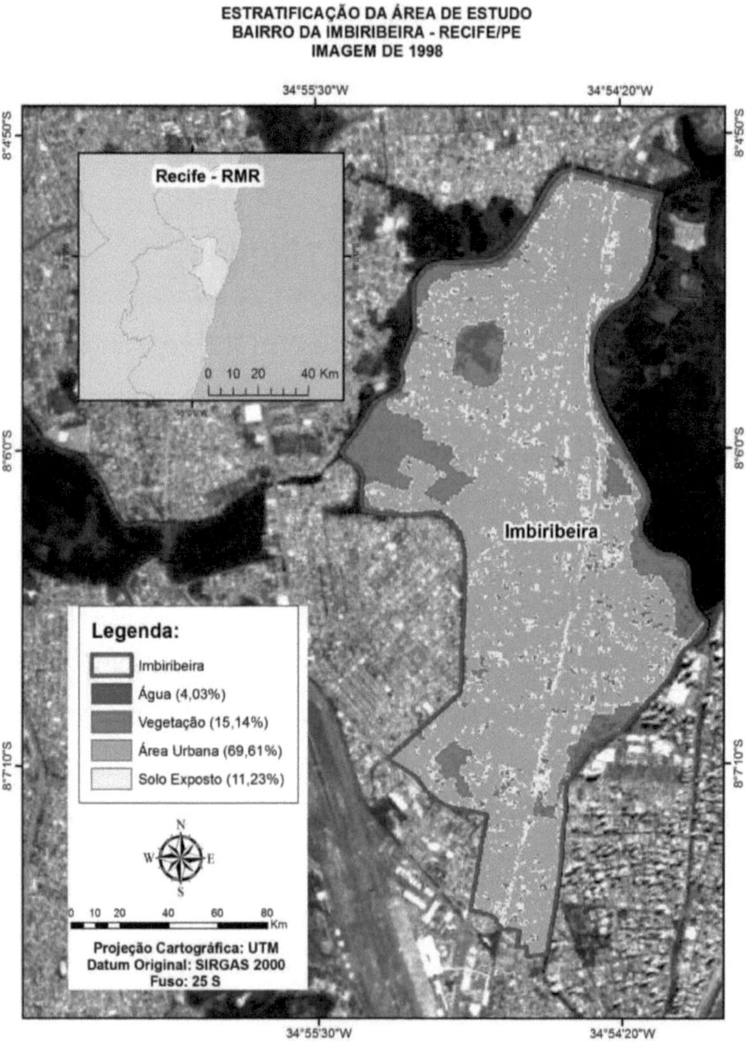

Figure 15 - Imbiribeira neighbourhood with 2016 image
Source: IBGE and INPE, Elaboration: Author

Table 5 - Percentage Stratification of the Imbiribeira Neighbourhood from 1984-2016

Stratification	1984	1998	2008	2016
Exposed Soil	9,10 %	6,42 %	0,00 %	11,23 %
Water	2,83 %	1,56 %	2,24 %	4,03%
Urban Area	62,34 %	69,09%	74,69%	69,61 %
Vegetation	25,72 %	22,93%	23,07%	15,14 %
Grand Total	**100 %**	**100%**	**100%**	**100%**

This image was the one with the best resolution and was taken from Google Earth, but without any georeferencing. As with the other images, a georeferencing process was necessary in order to stratify the image in an unsupervised process. The differentiation of the layers was obtained due to its high resolution in relation to the previous images. As a result, there is a higher percentage of exposed soil (11.23%) and water (4.03%), while vegetation (15.14%) has suffered from the increase in population in the neighbourhood and has declined sharply.

Some vegetation strips were replaced by exposed soil, but in the other images it was not detected or did not exist, likewise the urban area due to the better resolution decreased and also gave space to exposed soil referring to dirt roads, it is also verified that the araçá lagoon is a good part as if it were vegetation, due to the aerobic process of decomposition of the organic materials contained in the lagoon dumped by the population irregularly, thus a greenish carpet appears in the image causing its reflectance to be confused with vegetation as shown in the figure below:

Figure 16 - Araçá Lagoon with Reflectance Level Close to that of Vegetation
Source: Google Earth

CHAPTER 6

FINAL CONSIDERATIONS

It was possible to verify that in small areas it is difficult to stratify the layers of the images, but even so, by approximately extrapolating the reflectance values, it is possible to measure the ranges specified in the research. As expected, the urban area gradually increased as the years went by, due to natural urban expansion, the same occurred in exposed soils, known as vacant lots that were gradually occupied due to the need for growth and territorial occupation, One detail caught my attention: all the images downloaded from the Instituto de Pesquisas Espaciais website were not orthorectified, a good number of them were discarded due to unsuitable bands that didn't provide the information needed to carry out the digital image processing test, while others were discarded due to the main problem faced in the state when it comes to obtaining quality images: clouds, clouds are a constant problem, especially in the state on the Borborema plateau, from the forest zone onwards, where the most robust cloud condensations begin, preventing many studies due to this unavailability of quality images in terms of cloud formations, another problem faced was the use of ArcGIS software to obtain unsupervised Digital Image Processing, using the IsoData method, Due to the bands obtained, it was necessary to separate them into several layers of stratification, as the reference values were very close (Water/Vegetation) and the pixels had a total area of 900 m^2 . This is not considered sufficient resolution for a neighbourhood study to define classes, as the values of each pixel can be mixed with several characteristics, thus distorting the final result, making it an arduous task to obtain this stratification in the images used. At the end of the work, it can be seen that the development of large urban centres is underpinned by the shoulders of vegetation that already carries a burden of environmental protection and water bodies. The areas of exposed soil are also used because they have no defined use, As a general rule, the bodies of water have not undergone any changes in their composition, which would only be noticeable due to the silting up of the river or the filling in of a wetland with no apparent vegetation.

REFERENCES

Orbital Image Acquisition, INPE Available at <www.dgi.inpe.br/CDSR/>. Accessed on 02/01/2017.

BARCELLOS, Frederico C.; OLIVEIRA, Sônia Maria M.C. "New Sources of Data on Environmental Risks and Social Vulnerability". IBGE, Rio de Janeiro, 2008. p. 1-15.

Brasil Escola "History of Recife" <http://cidadebrasileira.brasilescola.uol.com.br/ pernambuco/historia-recife.htm>. Accessed on 20/01/2017.

FRANCISCO, Wagner de Cerqueria e. "Population growth in the world"; Brasil Escola. Available at <http://brasilescola.uol.com.br/geografia/o-crescimento-populacional- no-mundo.htm>. Accessed on 28 December 2016.

GROSTEIN, Marta Dora, Metropolis and Urban Expansion: the persistence of **"unsustainable"** processes**, 2001**
"Imbiribeira. Available at <https://pt.wikipedia.org/wiki/Imbiribeira>, Accessed on 09/01/17.

INMET "Periods of Highest and Lowest Temperatures and Climatological Rainfall".
Available at
<http://www.inmet.gov.br/portal/index.php?r=clima/mesTempo>, Visited on 02 January 2017.

Introduction to digital image processing / IBGE, First Geosciences Division of the Northeast. - Rio de Janeiro : IBGE, 2000.

Introduction to Remote Sensing Image Processing / Paulo Roberto Meneses Tati de Almeida - Brasília, Universidade de Brasília - UnB, 2012.
Law on the National System of Nature Conservation Units N° 9.985, of 18 July 2000.

Lei-uso e ocupação do solo-16176-1996-Recife-PE-consolidada-[19-06-2013].

MARTIN, Laurent "Satellite Image Orthorectification". Available at <http://www.engesat.com. br/ortorretificacao-o-que-e/>. Accessed on 16 January 2017.

Ministry of the Environment "Impact on Biodiversity". Available at <http://www.mma.gov.br/biodiversidade/biodiversidade-global/impactos> Visited on 07 January 2017.

Digital processing of remote sensing images / Alvaro Penteado Crósta. - rev. ed. - Campinas, SP:IG/UNICAMP, © 1992.

ROSE, Ricardo Ernesto , Debates Culturais "Brazil's Urban Growth" <http://www.debatesculturais.com.br/o-crescimento-urbano-brasileiro/> Visited 28 December 2017

Department of the Environment, Recife Master Plan "Municipal Conservation Units" - Recife: SEMA, 2012.

Visit Recife " History of Recife: From its emergence to the present day". Available at <http://visitarecife. com.br/historia-do-recife/> Accessed on 20 January 2017.

MIX
Papier aus verantwortungsvollen Quellen
Paper from responsible sources
FSC® C105338

Printed by Books on Demand GmbH, Norderstedt / Germany